THE HUMAN BODY

BREAKTHROUGHS IN SCIENCE

THE HUMAN BODY

CAROL J. AMATO

ILLUSTRATIONS BY STEVEN MOROS

SMITHMARK

DEDICATION

■ ■ ■ ■ ■

To all the tireless researchers looking for cures to cancer and AIDS
and all the other diseases afflicting humankind.

ACKNOWLEDGEMENTS

■ ■ ■ ■ ■

I would like to thank Biomagnetic Technologies, Inc., San Diego,
California, and Frank Forestieri, of FJF Communications, New York,
New York, for their information on the biomagnetometer (MEG); and
Medtronics, Inc., Fridley, Minnesota, and Telectronics, Denver,
Colorado, for their information on pacemakers.

A FRIEDMAN GROUP BOOK

This edition published in 1992
by SMITHMARK Publishers Inc.
112 Madison Avenue
New York, New York 10016

ISBN 0-8317-1014-4

BREAKTHROUGHS IN SCIENCE: THE HUMAN BODY
was prepared and produced by
Michael Friedman Publishing Group, Inc.
15 West 26th Street
New York, New York 10010

Editor: Dana Rosen
Art Director: Jeff Batzli
Designer: Lynne Yeamans
Photography Researcher: Daniella Jo Nilva
Illustrator: Steven Moros

Typeset by Bookworks Plus
Color separations by Rainbow Graphic Arts Co.
Printed and bound in Hong Kong by Leefung-Asco Printers Ltd.

SMITHMARK Books are available for bulk purchase for sales promotions and premium use.
For details write or telephone the Manager of Special Sales, SMITHMARK Publishers Inc.,
112 Madison Avenue, New York, New York 10016. (212) 532-6600.

TABLE OF CONTENTS

INTRODUCTION

◻ ◻ ◻ ◻ ◻

For thousands of years, humans have wondered what makes us sick and what makes us well. Ancient cultures, including those of the Egyptians, the Babylonians, the Minoans, and the Indians, all had developed the science of medicine and the art of healing. Though it would be very difficult for us to imagine being healthy in a world without antibiotics, chemotherapy, or the Salk polio vaccine, we are now beginning to realize our ancestors knew more about health than we thought. They practiced healing with herbs and chants, and until recently, we considered such methods primitive and superstitious. In this century, we have rediscovered some of their secrets.

Chinese acupuncture, the art of placing needles in specific places in the body to relieve pain, which has been practiced for thousands of years, has now gained wide acceptance and is in use around the world. Holistic doctors, doctors who consider the mind and body as one, are using herbs in the treatment of disease. Because much of what we call modern medicine is based on historical methods passed on through the ages, any discussion of great medical breakthroughs should include the discoveries made by these ancient people.

For instance, the ancient Egyptians played a major part in turning medicine into a science. They learned the art of mummification—the practice of wrapping bodies in cloth to preserve them for long periods of time. The mummies found today, 5,000 years after they were buried, are still intact.

The ancient Greeks rid medicine of magic and superstition by adopting a totally scientific approach. Hippocrates (Hip-*pah*-cruh-teez) (c460-c360 B.C.), a famous Greek doctor, knew that fresh air, rest, and a good diet were essential to recovery from disease. Hippocrates wrote many books and speeches on medicine, and he is called the "Father of Medicine." His oath, in which doctors promise to do everything they can to save lives, is still sworn to today by every doctor graduating from medical school.

Ancient Egyptians practiced the art of mummification. This mummy case contains a body that has been wrapped in cloth and mummified, to preserve it for a long time. The mummy case has a likeness of the dead person carved on the top.

BREAKTHROUGHS IN ANATOMY, PHYSIOLOGY, AND PSYCHOLOGY

BREAKTHROUGHS IN ANATOMY

Anatomy, or the structure of the human body, has fascinated humanity for centuries.

In prehistoric times, those who were the most skillful at caring for broken bones and tending wounds became the first medicine men and women, or the world's first doctors. By 3,500 B.C., about 5,000 years ago, the practice of medicine was well established. Historical records show that in ancient Egypt there were many physicians and surgeons. But like the early Stone Age healers, Egyptian doctors often depended on magic, the spirits, and the gods to help them find a cure.

The ancient Greeks made medicine a science. For the first time in history, the doctor carefully studied a patient's symptoms and did not rely so much on magic or the whims of the gods. Medical schools were set up throughout Greece.

When the Romans conquered the ancient world, the city of Rome became the center of the civilized world. Many Greek doctors went to Rome to practice medicine. One Greek, Galen, practiced medicine in Rome around A.D. 161. He followed many of the methods of Hippocrates from 600 years before, and the books that he wrote on anatomy were used as textbooks in medical schools for the next 1,300 years! But Galen had never dissected a human body. He had dissected dogs, goats, and monkeys, and had assumed that human anatomy was the same. As a result, many of his ideas about how the human body worked were wrong.

In the 1500s, however, a young and brilliant Flemish medical student, Andreas Vesalius (1514–1564), gave doctors a true picture of human anatomy. For years, human dissections were not allowed, but by Vesalius's time, medical schools were permitted to dissect the bodies of dead

This bust of Hippocrates, a famous Greek doctor, was found on the Isle of Cos, the location of a famous fifth century B.C. medical center. Hippocrates, called the ''Father of Medicine,'' conceived the ''Hippocratic Oath,'' which is still sworn to by every doctor graduating from medical school.

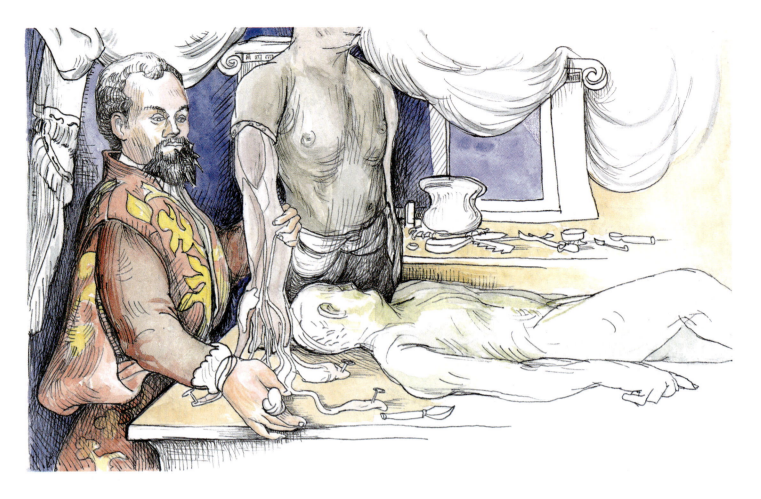

■ ■ ■ ■ ■ ■ ■ ■ ■ ■ ■ ■ ■

Flemish medical student Andreas Vesalius did not believe that doctors could learn about the human body by dissecting animals. He practiced dissection on dead criminals and learned that earlier guesswork on the anatomy of the human body had been completely wrong.

criminals. During a dissection, however, teachers and students followed along according to Galen's book. Cutting up a human body according to an animal's anatomy made these dissections careless and inaccurate. Vesalius, who had long been a student of anatomy, discovered Galen's mistakes when a book publisher in Italy asked him to update Galen's work. As a result, Vesalius wrote a new book on anatomy to replace Galen's.

Fortunately, many of the great Italian artists of the day, like Leonardo da Vinci, Michelangelo, and Raphael, were also interested in anatomy. These artists believed that if they learned how the human body worked and moved, they would be able to draw better paintings and create better sculptures. Vesalius asked Stephen Calcar, a student of the great artist Titian, to do the illustrations for his new book on anatomy.

Published in 1543, *De Humani Corporis Fabrica*, which means *Concerning the Structure of the Human Body*, changed the course of medical science. Vesalius's book was the first complete and accurate anatomy textbook in the world and later earned Vesalius the title "Father of Anatomy."

BREAKTHROUGHS IN PHYSIOLOGY

While anatomy is the study of the structure of the human body, physiology is the science of how the body works or functions.

■ Blood Chemistry and Circulation

Although English doctor William Harvey (1578–1657) is known for discovering how blood moves through the body, many of our ancient ancestors understood how blood circulates thousands of years before Harvey was born. The ancient Egyptians described blood circulation well, but the ancient Sumerians of the Tigris-Euphrates Valley mistakenly believed the liver pumped the blood. A wise Chinese Emperor, Huang-ti, who lived over 4,000 years ago, clearly had an excellent understanding about the flow of blood through the body. In his book, *Nei Ching (The Book of Medicine)*, he wrote: "All the blood in the body is under the control of the heart . . . the blood current flows continuously in a circle and never stops."

Yet, William Harvey is considered the "Father of Physiology." Harvey was a successful doctor who served as physician for King James I and King Charles I of England. Like Vesalius, Harvey studied medicine at the University of Padua in Italy. Harvey became interested in the beating of the heart and in the movement of the blood. His teacher, Girolamo Fabrizio, discovered that there were valves,

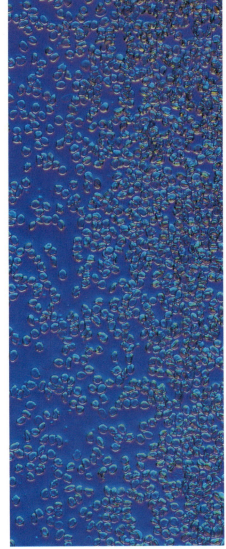

© Brian Parker/Tom Stack & Associates

Blood flows through the arteries and veins of the body. Although it appears red when it comes into contact with air, such as when a person has a cut, it is normally blue. This is why veins look blue. Blood is composed of a yellowish fluid called plasma, which contains millions of cells. A cubic millimeter of blood has about 5,000,000 red corpuscles (or red blood cells, which carry oxygen through the body); 5,000 to 10,000 white corpuscles (cells that help fight disease); and 200,000 to 300,000 platelets (which help form blood clots in order to heal injuries).

• • • • • • • • • • • • •

Opposite page: This illustration shows how the blood circulates through the heart. The heart, a hollow muscle, receives blood from the veins and sends it into and through the arteries. The red blood has oxygen mixed with it; the blue blood does not. **Inset:** This close-up shows blood moving through the capillaries in the lungs. As it does so, it picks up oxygen.

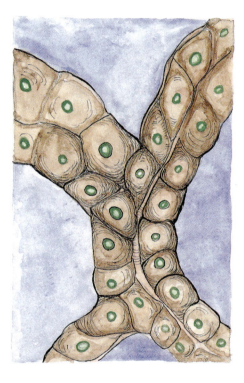

• • • • • • • • • • • • •

This illustration of lung tissue shows cells like those Marcello Malpighi saw under the miscroscope. Because of his discovery that living organisms were made of many different types of tissue, several body parts bear his name. These include the Malpighian tufts in the kidneys, the Malpighian layer in the epidermis (skin), and the Malpighian corpuscles in the spleen.

which he called "little doors," in the veins around the heart. Another doctor who lived in Spain, Michael Servetus (1511–1553), developed the idea that the blood flowed from the lungs to the heart.

Harvey, however, realized that when the blood passed through these valves, it flowed in only one direction— toward the heart, but never away from it. Although he wasn't the first person to talk about circulation this way, in 1628 Harvey published *An Anatomical Treatise on the Motion of the Heart and Blood in Animals*, which popularized his ideas about the endless circulation of blood.

◼ The Discovery of Blood Types

Almost 300 years after Harvey's observations about blood circulation, Karl Landsteiner (1868–1943) of Vienna, Austria, discovered that blood came in different types.

In 1901, he developed the modern classification of the four main types of blood, *A, B, AB,* and *O.* His discovery led to the safe use of blood transfusions from one person to another. In 1930, Landsteiner received the Nobel prize in medicine and physiology.

◼ The Discovery of Cells

All the living parts of the body are made up of cells of many different kinds, shapes, and sizes. Brain cells differ from the cells of the skin, bones, or muscles. Cells are the fundamental "building blocks" of living matter.

Around 1600, Italian Marcello Malpighi (Mar-*chel*-lo Mahl-*pee*-ghee), one of the first professors of anatomy to use the newly invented microscope, discovered that living organisms were made up of different types of tissue. In turn, each tissue was composed of many parts, visible only under the microscope. He called these parts "saccules," and we now call them cells.

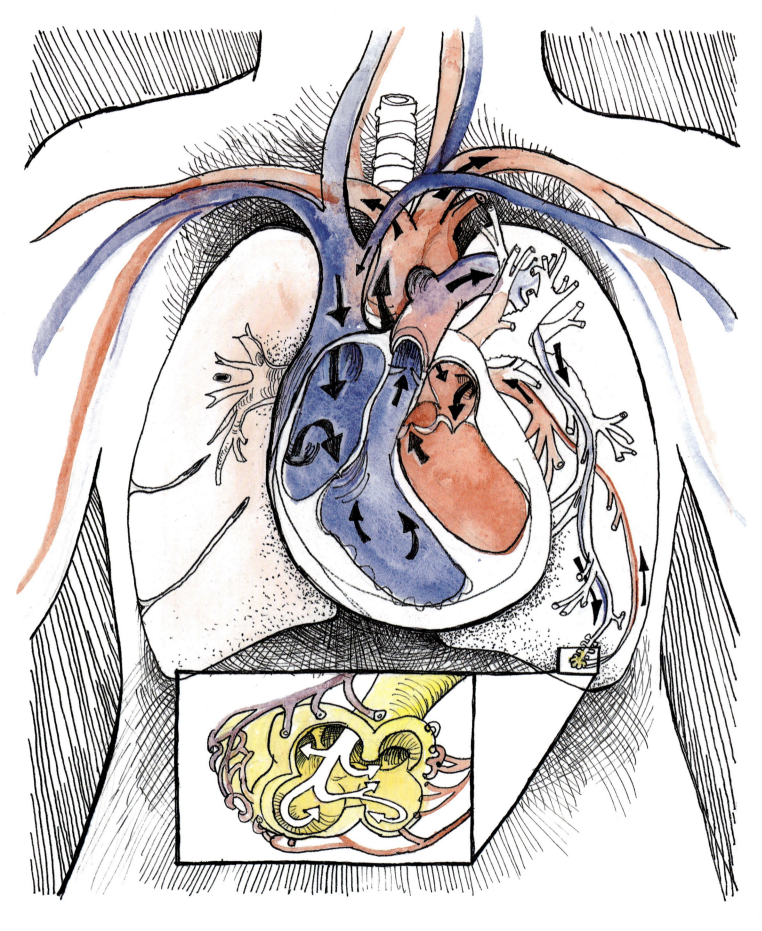

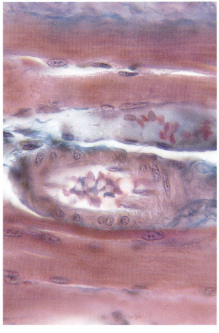

This photograph shows the muscles and blood vessels of an animal tongue as seen through a microscope.

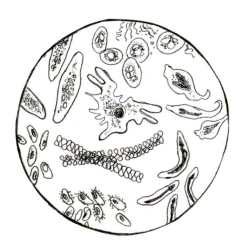

When Antonie van Leeuwenhoek studied stagnant water under the microscope, he discovered many one-celled animals we now call amoebas and protozoa. He also discovered bacteria, or germs.

Malpighi founded the science of microscopic anatomy. By studying layers of skin and lung tissue, he also discovered the capillaries, the tiny, fine blood vessels that connect arteries to veins.

■ The Discovery of Germs

During the 1600s, an amateur scientist from Holland, Antonie van Leeuwenhoek (Lee-ven-hoke) (1632–1723), was the first person to see germs. He was very interested in grinding lenses and making magnifying glasses. He made small microscopes at first, but was soon making some of the best microscopes in the world.

Van Leeuwenhoek looked at many things under the microscope, including insects such as ants, spiders, and bees. He examined blood, bones, eyes, hair, and muscles. When he looked at a drop of stagnant water, he was shocked to discover little creatures swimming around. These are one-celled animals we call protozoa, and included amoebas and paramecia. He also saw bacteria, which he called "animalcules," but he didn't realize these germs cause disease.

Van Leeuwenhoek made many other discoveries in addition to his discovery of germs. While studying the transparent tail of an eel, he concluded that the vessels leading away from the heart were arteries and those going toward the heart were veins.

In 1674, he discovered red corpuscles in the blood. Red corpuscles carry oxygen through the body. Van Leeuwenhoek reported all of his findings to the Royal Society of London. British scientists were so impressed by his work, they made van Leeuwenhoek a member of the society.

In 1716, when he was 84, van Leeuwenhoek received an honorary degree from the University of Louvain (Loo-vanh). His work laid the foundation for that of Louis Pasteur (*see Chapter Two*) more than a hundred years later.

BREAKTHROUGHS IN PSYCHOLOGY

Until recently, many doctors believed that the study of the mind had little to do with the study of the body.

We now know, however, that people become mentally ill much in the same way as they become physically ill. The mind and the body are linked, and the mental state of a person can influence his or her health. It also makes a difference in how well a patient recovers from a physical ailment. We now realize, too, that people can recover from many types of mental illnesses.

For centuries, though, people suffering from mental illness were horribly mistreated. They were chained and locked in dungeons, where they had no hope of recovery. Many believed the mentally ill were possessed by the devil.

In past centuries, people believed that the mentally ill were possessed by the devil and so chained and locked them in dungeons. Because they received no treatment of any kind, there was no way for them to get well.

In the late 1700s, however, a French doctor named Philippe Pinel (Pin-*el*) (1745–1826) became one of the first to realize that the mentally ill needed proper treatment. Pinel didn't believe that mental patients were possessed by the devil. Instead, he insisted that mentally ill people be treated with kindness and understanding, fresh air, exercise, and good food. He tried to learn the source of his patients' problems in order to cure them. Pinel is known as the "Father of Psychiatry."

In the United States, Dorothea Dix (1802–1887) was the first to help mental patients much in the same way as Pinel helped the mentally ill in France. Dix didn't believe in putting mental patients in prison with criminals. She helped

■ ■ ■ ■ ■ ■ ■ ■ ■ ■ ■

Austrian doctor Sigmund Freud believed that mental problems were caused by experiences in early childhood. This illustration shows him in his office ''psychoanalyzing'' a patient. The patient is lying on the couch to help him relax, and he is telling Freud all about his problems in hopes of understanding and overcoming them.

start mental hospitals in fifteen states of the United States, and in Canada, Europe, and Japan as well. Though the mentally ill were now in hospitals, they were still not offered proper treatment.

Despite the efforts of reformers like Pinel and Dix, the causes of mental illness received little scientific attention until an Austrian doctor named Sigmund Freud (1856–1939) began to study the nature of the human mind. Freud believed that mental problems were caused by negative experiences in early childhood. Known as the "Father of Psychology," Freud invented psychoanalysis, a treatment method in which people talk about their problems in order to understand and overcome them.

Freud learned more about the human mind in half of his lifetime than the world could figure out in all the centuries before him. Today, however, in some circles, many aspects of Freud's work have fallen into disfavor. Freud attributed the causes of all mental problems to early childhood events. We now know, however, that mental disorders can arise for many different reasons. Chemical imbalances in the brain and body can cause many mental diseases, such as schizophrenia, but we are only now beginning research on these problems.

Some of the old, inaccurate attitudes about the mentally ill are with us today. While we realize that people can fully recover from physical illnesses, many find it difficult to accept that people can recover from mental illnesses. They believe that once a person has been mentally ill, his or her mind is permanently weakened. Knowledge of the treatment of mental illness can prevent a person from getting a job or holding political office.

It is to be hoped that our attitudes toward those recovering from mental illness will one day match our healthier attitudes toward those recovering from physical illness.

BREAKTHROUGHS IN MEDICINES

The art of healing has been practiced since prehistoric times. Our early ancestors understood much about medicines. The ancient Chinese, for example, gathered a great deal of knowledge about drugs used for healing. Emperor Shen Nung, China's first physician, who lived 5,000 years ago, prescribed ephedrine, a drug made from a local plant, as a hay fever and asthma medication. It is still used today.

The ancient Jews recognized that certain foods cause illness, and they banned the foods that made people sick. They also kept sick and well people apart, thus becoming the first to invent ways to prevent the spread of disease.

This nineteenth-century pharmacy in Canton has a typical open front. The Chinese have long been experts in the art of healing and have used drugs as medicine for over 5,000 years.

In ancient India, doctors discovered that rats carried bubonic plague and mosquitoes carried malaria, although they didn't realize that it was the germs these creatures carried, and not the animals themselves, which were responsible for the diseases.

The Romans built aqueducts to carry fresh water through their cities, for they knew that sewage and dirty drinking water caused epidemics. They also built some of the first hospitals, initially to help injured soldiers, but later on to aid the Roman people.

This is a Roman aqueduct that dates from the early first century. The Romans used aqueducts to carry fresh water through their cities.

THE DEVELOPMENT OF ANESTHETICS

Imagine having a doctor perform an operation or having a dentist drill your teeth without giving you a painkiller first. Before 1800, operations were performed without painkillers or anesthetics. Doctors gave their patients whiskey or opium, but these really didn't work.

In 1800, an Englishman, Sir Humphry Davy, suggested that nitrous oxide (commonly called "laughing gas" because it made patients laugh and act silly) be used to stop pain. No one paid attention to his suggestion until the 1840s, when Dr. Crawford W. Long, an American surgeon from Georgia, came on the scene. Long realized that people who sniffed a gas called "ether" seemed to feel no pain when they bumped into things. He began using ether to anesthetize his patients during operations. Long, however, didn't publish his discovery until 1849.

In 1844, a dentist named Horace Wells made a similar observation about people who inhaled laughing gas. Wells began using the gas in his dentistry. In 1846, Dr. Charles T. Jackson suggested the use of ether to Dr. William T. G. Morton, a dentist who had studied under Wells. Morton used ether in dental surgery. He also administered the gas to a young woman who was having her leg amputated, and she did not feel any pain during the operation.

■ ■ ■ ■ ■ ■ ■ ■ ■ ■ ■ ■

Horace Wells, a dentist from Hartford, Connecticut, used laughing gas as an anesthetic in his practice.

■ ■ ■ ■ ■ ■ ■ ■ ■ ■ ■ ■

Opposite page: Englishman Alexander Fleming accidentally discovered penicillin when he noticed that a mold growing in a dish of bacteria had killed the germs closest to it.

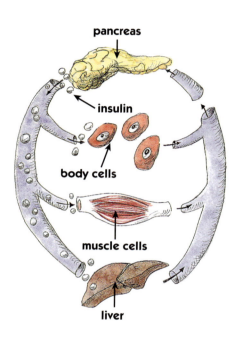

pancreas

insulin

body cells

muscle cells

liver

■ ■ ■ ■ ■ ■ ■ ■ ■ ■ ■ ■

When the body takes in carbo-hydrates, enzymes convert them to glucose, or sugars. These sugars are sent to the liver, where they are converted to glycogen. The glycogen releases free glucose into the blood as energy when the body needs it. Insulin, which is produced in the pancreas, helps the muscle cells and other body cells take in the glucose. Insulin also helps regulate the amount of glucose taken into the cells. Diabetic people do not produce insulin, so they must take insulin shots to help their bodies use glucose properly.

In 1847, Sir James Simpson, a Scottish doctor, first used another gas, chloroform, as an anesthetic. These experiments made the use of ether, laughing gas, and chloroform popular in the United States and throughout Europe.

Anesthetics made many kinds of painful, but necessary, surgeries possible.

THE DISCOVERY OF INSULIN

There are dozens of organs in the body called glands, which make and distribute chemicals the body requires. Diabetes is a disease in which the pancreas, a large gland located behind the lower part of the stomach, cannot properly convert sugar into energy. Too much sugar builds up in the blood and can cause kidney and heart disease, blindness, and even death.

Before 1921, most victims of diabetes died. By 1921, however, two Canadian doctors, Sir Frederick Grant Banting and Charles Herbert Best, perfected insulin. Insulin is a hormone substance that acts as a chemical messenger by helping the pancreas process the blood sugar.

Insulin does not cure diabetes, but it helps to control the disease. Some diabetics must use insulin every day.

THE DISCOVERY OF PENICILLIN

Until 1928, doctors had no way of treating patients with bacteria-caused diseases like influenza or strep throat. The cure for many of these diseases was discovered by accident. At a hospital in London, the researcher named Alexander Fleming was growing staph germs (a type of bacteria) in a dish for an experiment. One day, he noticed a strange mold growing in the dish. He also noticed that the staph germs closest to the mold had been killed.

Polish refugee Marie Curie (1867–1934) worked for many long years in her laboratory to isolate the radioactive elements in the ore pitchblende. Her work helped bring about radiation therapy for cancer.

Fleming discovered that the mold was part of the penicillin family. This mold is commonly found in moldy bread. The Egyptians used the mold from old bread to treat wounds. Yet, not until this century did modern doctors discover that antibiotics—drugs used to treat infections—could be made from mold. Fleming separated the mold that killed the staph germs and named it "penicillin." Penicillin was produced for widespread use by 1944. This substance launched the era of antibiotics.

Although penicillin was considered the new miracle drug, it couldn't cure certain kinds of infections. In New Jersey, a Rutgers University scientist, Dr. Selman A. Waksman, discovered a microbe, *Streptomyces griseus*, in some dirt. Many years later, he discovered this microbe again. He called the new antibiotic he developed from the microbe, streptomycin. Unlike penicillin, streptomycin proved very effective against tuberculosis and urinary tract infections. Since then, a wide variety of antibiotics have been developed to treat many different infections.

THE DISCOVERY OF RADIUM

Madame Marie Curie was a Polish refugee, who along with her French husband, Pierre Curie, set out to isolate the radioactive elements in the mineral pitchblende. They chose pitchblende because they found that it was more radioactive than any other substance they had studied. In 1898, after years of exhaustive work, Marie Curie discovered two elements, one which she named *polonium* after her native country, Poland, and the other *radium*. The discovery of radium is significant because it gives off a powerful ray that is able to help in destroying cancerous cells.

Radium was used to treat cancer until a metal substance called Cobalt 60 was found to be more effective. Work on cancer-destroying rays didn't stop with Marie and Pierre Curie. Their daughter, Irene, and her husband, Frederick Joliot, carried on the Curies' work, playing a vital role in the discovery of radioactive isotopes, elements which have similar chemical properties, but differ in behavior.

The discovery of radium earned Marie and Pierre Curie a Nobel prize for physics in 1903.

THE DISCOVERY OF VACCINES

In past centuries, measles and many other diseases caused by bacteria were fatal. Sometimes whole families were wiped out by these epidemics. Measles, which often strikes children, but can also infect adults, causes high fever and red spots all over the body. John F. Enders, who received the Nobel prize for medicine in 1954, produced the measles vaccine in 1962. Almost everyone has received a vaccine—the shot in the arm or leg that helps the body fight off these once deadly diseases.

Diphtheria is a disease that causes a false membrane to form in the air passages. Characterized by high fever and inflammation of the heart, this disease can be fatal. The diphtheria virus was discovered in 1884 by a German bacteriologist named Friedrich Löffler (1852–1915). In 1894, Emil von Behring (1854–1917), another German bacteriologist, produced a vaccine against the dreaded disease.

Max Theiler won a Nobel prize for discovering a vaccine for yellow fever. Transmitted by the bite of a mosquito, yellow fever causes jaundice (a liver disease that turns the skin yellow), vomiting, and collapse of the liver.

Despite these breakthroughs in medicines, many people still suffer from these diseases because they are unaware of the available vaccines. This lack of knowledge has prevented these diseases from being completely eliminated.

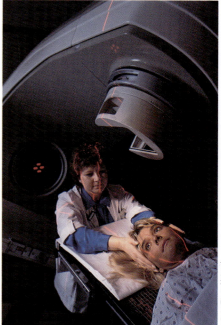

© Duka/Photo Network

This patient is undergoing radiation treatment. Radiation, made possible by Madame Curie's discovery of radium, is used to kill cancerous cells.

■ Smallpox Vaccine

The dreaded smallpox disease affected some sixty million people throughout Europe during the eighteenth century. Many people who caught smallpox died. Those who did not die were blinded or were badly scarred from the rash the disease produced.

In the early 1700s, an Englishwoman, Lady Mary Wortley Montagu, saw Turkish people giving inoculations consisting of fluid taken from smallpox sores to their children. Suggesting this treatment of the disease in her own

■ ■ ■ ■ ■ ■ ■ ■ ■ ■ ■ ■ ■ ■ ■

This illustration shows Dr. Jenner scraping cells from a cow to test his theory that contracting cowpox makes people immune to smallpox.

country, she and her children were among the first of thousands in England to be so injected. The method was risky, however, for the patient could get too strong a dose of the smallpox fluid and die.

In England, in 1796, Dr. Edward Jenner learned that farmers were touching infected cows and contracting cowpox, a milder form of smallpox. Once they recovered from cowpox, the farmers seemed immune to smallpox. Jenner tested his theories and discovered the vaccine for smallpox. Nearly two hundred years passed before enough people were vaccinated so the disease could be wiped out.

◾ Polio Vaccine

A disease called poliomyelitis, or polio, crippled and killed many people by spreading from the blood into the nerve cells of the brain or spinal cord. Many adults forty and fifty years of age today can recall a time when they were not allowed to swim during the summer for fear of catching polio. People believed that the polio virus spread through water when the temperatures were high.

Children and adults alike contracted polio. A late president of the United States, Franklin Delano Roosevelt, was a victim as a young man. Because of polio, Roosevelt was unable to walk without the aid of crutches.

At the University of Pittsburgh, a young medical scientist named Dr. Jonas E. Salk (1914–   ) developed a vaccine against polio by growing three types of viruses, killing them, and then adding them together. The Salk vaccine was first tested on monkeys, and then, in 1953, on people in Pittsburgh, including Salk's own children. It then began to be manufactured by five drug companies. In 1954, 440,000 children in the United States received injections of the Salk vaccine. By 1955, the world realized that polio had been conquered.

In 1953, Dr. Jonas Salk discovered a vaccine for polio, a crippling disease. Today, Dr. Salk is working on a vaccine for AIDS, a disease of the immune system. Inset: This close-up shows how the polio virus looked to Jonas Salk when he viewed it under a microscope.

In 1960, Dr. Albert Sabin developed an oral form of the polio vaccine. The vaccine was coated on the outside with a sugar cube. Children no longer had to have shots to be safe from polio, and the oral vaccine offered longer-lasting protection.

◼ Rabies Vaccine

One of the most feared diseases in the world is rabies, or hydrophobia. This infection of the central nervous system is caused by a virus transmitted through the bite of animals stricken with rabies, which may include skunks, foxes, bats, raccoons, dogs, and cats. Before the discovery of the vaccine, people who caught rabies almost always died.

In 1884, Louis Pasteur (1822–1895), a French chemist and microbiologist, discovered the rabies virus in the spinal cord, brain, and saliva of infected animals. After these animals died, Pasteur dried their spinal cords and used them to make a serum. He injected the fluid into other animals, some of which were healthy and others that were in the first stages of the illness. He soon discovered that if he gave the animals the injections every day for fourteen to thirty days, depending on how sick the animal was, the well ones did not become sick and the sick ones did not get any worse. In fact, they eventually recovered. Had he discovered a cure for rabies?

Word spread about Pasteur's experiments, and before his testing was done, a frantic mother brought her young son to him in desperation. A rabid dog had bitten the boy, and the mother was certain he would soon die. Pasteur had never tested his vaccine on humans, but because he knew the mother was right, he felt he must risk giving it to the boy. The boy became ill, but the injections kept him from dying. Two months later, the boy was completely well. Pasteur had indeed discovered a cure for rabies.

3

BREAKTHROUGHS IN MEDICAL TECHNOLOGY

Through the ages, medical inventions have helped doctors save lives. Who were the people who came up with these ideas? How did they think of them? Let's take a look.

THE MICROSCOPE

For centuries, scientists had a hard time learning about the human body because they could not see many of its parts. The microscope, a device that allows us to see enlarged pictures of very small objects, solved many mysteries for scientists and opened up whole new areas of research.

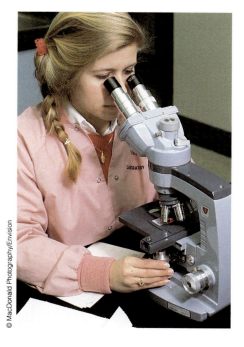

Above: A lab technician studies blood samples under a modern microscope. At left: These are red blood cells as seen under a microscope.

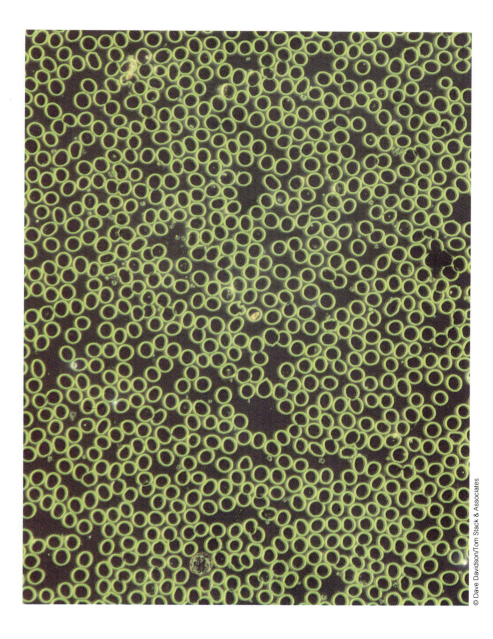

Antonie van Leeuwenhoek was a dry goods store owner who loved looking at all kinds of objects under the microscopes he built. When he looked at pond water under the lens, he discovered little creatures swimming around. These were amoebas and paramecia, and his work helped pave the way for the science of microbiology.

At left: A lab technician studies leukemia cells through an electron microscope.

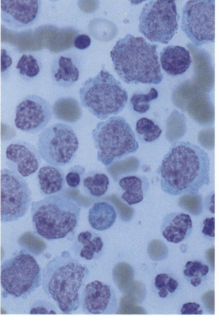

Above: This shows how leukemia cells look when viewed through a microscope.

Zacharias Janssen of the Netherlands is credited with inventing the first microscope in 1590. Unfortunately, his microscopes didn't work very well. It was Antonie van Leeuwenhoek (*see Chapter One*) who invented the best microscopes. Van Leeuwenhoek was not formally educated. He was a dry goods store owner, not a scientist, but he enjoyed looking at cloth, coins, bees, flies, blood, and water under lenses he ground himself. No one knows how he learned to grind lenses, but by 1660, he was grinding lenses used to make the best microscopes of his day.

Almost everyone has used a simple microscope in science class. But there are many different kinds of microscopes that are much more complex. The electron microscope, for example, uses an electron beam to "light up" the specimen being studied, thereby allowing much greater magnification.

© Steven Mark Needham/Envision

This broken thermometer has let the mercury float free. As you can see, mercury does not run like water when it escapes, but rather, forms little beads.

THE CLINICAL THERMOMETER

When people are ill, their body temperature often changes. If it gets too high or too low, they can die. Before the thermometer was invented, doctors could not tell how high a person's fever was or how cold the body might become.

In 1590, the same year that Janssen invented the microscope, the Italian mathematician Galileo Galilei invented the thermometer. Best known for his invention of the telescope and for his discovery that the moon has mountains, Galileo invented the thermometer by accident.

Galileo's thermometer consisted of a long, straight, glass tube, with one end open and the other end shaped like a bulb. Galileo put the open end into some water, then held the bulb end in his hand. Soon, the heat from his hand made the air in the tube heat up and expand. Then, the water began to bubble as the warm air escaped into it. When Galileo took his hand away, the air in the tube began to cool off and contract. The water stopped bubbling and flowed up into the tube.

Galileo discovered a simple way to measure temperature, but he didn't realize that his device could be used in medicine. However, a friend of Galileo's, a doctor named Sanctorius, wondered about the thermometer's possible medical use. He believed Galileo's invention would be valuable in monitoring the progress of a person's illness.

Sanctorius began to build his own thermometer. He used the open end and bulb end from Galileo's design, but instead of using a straight tube, he shaped it into a series of loops. The open end of the tube still went into the water, but the bulb end went into the patient's mouth. For twenty minutes, during which the patient must have been very uncomfortable, Sanctorius counted the bubbles. The higher the temperature was, the more bubbles there were.

Sanctorius also watched how far into the tube the cooling and contracting air drew the water. He noticed that the water was always at the same level for healthy people. As a result, Sanctorius calculated the height of a fever by comparing its water level against the level for healthy people.

Sanctorius made three more improved models of his thermometer, but each one shared the same problem: The glass tube and the water were exposed to the air, and so could not be totally accurate.

In 1631, French chemist Jean Rey created a model which reacted to body heat alone by sealing the water inside the

Galileo invented the first thermometer, but the great scientist saw no practical application for it. His friend, a doctor named Sanctorius, wanted to use Galileo's design to measure the temperature of sick people. He modified Galileo's straight-tube design into a series of loops. One end was open, which he put into a bowl of water. The other end had a bulb, which he put in the patient's mouth. How high the water rose into the tubes showed how high the patient's temperature was.

thermometer's tube. Later, alcohol replaced the water and the tube became much smaller.

In 1714, a German physicist named Gabriel Daniel Fahrenheit (1686–1736) made the first thermometer that used mercury instead of alcohol. With this thermometer, he created the temperature scale that is named after him.

Despite the change to mercury, doctors found that the thermometer would not keep the temperature once it was removed from the patient's mouth. To solve this problem, an English doctor, Sir Thomas Clifford Allbutt (1836–1926), designed a thermometer with a little "kink" near its base. As the person's body heat warmed the thermometer, the mercury rose past the kink. The kink then prevented the mercury from falling back down the tube unless the thermometer was shaken very hard. All clinical thermometers featured this little kink until a few years ago.

Today, disposable plastic thermometers have replaced glass ones. Plastic thermometers don't break, and they read a temperature in seconds rather than minutes. These thermometers change color at certain temperatures or have metallic coils that are affected by heat. Some types can simply be placed on the patient's forehead.

THE STETHOSCOPE

Until the 1800s, doctors listened to a patient's heartbeat by placing an ear against the patient's chest. Unfortunately, other body sounds got in the way of hearing the heartbeat clearly.

In 1817, however, a French doctor, René Laennec (Len-*nek*) (1781–1826), changed these outdated methods. Laennec suspected one of his patients had a heart problem. She was very heavy, however, which made hearing her heartbeat impossible.

Walking to the hospital the next morning, Laennec saw some children playing with a board. Some had their ears at one end, while another child tapped the other end of the board with a stick. Leannec saw that the children heard the sounds of the tapping traveling along the board.

Laennec returned to his patient with a new idea. Rolling a piece of paper into a tube, he put one end next to the patient's chest, and the other to his ear. He could hear every heartbeat!

Laennec created some wooden models of the tube and called it a stethoscope, from the Greek words for "the chest" (*stethos*) and "to observe" (*skopos*). This wooden stethoscope evolved into the models doctors use today.

René Laennec invented the stethoscope after seeing some children transmit tapping noises to each other via a wooden plank. He rolled up some papers and listened to a patient's chest through the roll.

THE X-RAY

▪▪▪

Before the X-ray, doctors guessed at how to reset broken bones or locate tumors. There was no way to see inside the patient's body.

Scientists knew that electricity was noisy when it traveled through air, but they wondered how it would react in a vacuum. They were surprised to discover that the flashes and crackles of electricity were replaced by glowing light.

An English scientist interested in this effect, Sir William Crookes (1832–1919), created a glass tube without air and with two terminals inside. One terminal was called an "anode," and the other a "cathode." When electricity passed

▪▪▪▪▪▪▪▪▪▪▪▪▪▪▪▪▪▪

German physicist Sir Wilhelm Conrad Roentgen focused the invisible rays from the Crookes Vacuum Tube onto a piece of cardboard treated with a light-sensitive substance and was shocked when he could see the bones of his hand!

through the tube, invisible rays traveled from the anode to the cathode, making the cathode glow. He called his invention the Crookes Vacuum Tube.

On November 8, 1895, a German physicist, Sir Wilhelm Conrad Roentgen (Rent-jen) (1845–1923), experimenting with the Crookes Vacuum Tube, discovered that even when the tube was covered with black cloth, the invisible rays still passed through. He put up a piece of cardboard painted with a light-sensitive chemical called barium platinocyanide. He blocked the rays with a piece of wood, and saw that a shadow of the wood continued to show up on the cardboard. When he put his hand in front of the rays, Roentgen was stunned to see not his hand, but the outline of the bones inside. The scientist knew he was seeing the first picture of the inside of the human body.

In January of 1896, Roentgen presented his new discovery to the Physical Medical Society of Wurzburg. He called the rays "X-rays," because he didn't know what they were. The "X" stood for the unknown factor.

The miraculous discovery of the X-ray means illnesses can be diagnosed more easily, but doctors have to be careful in using this powerful ray. Patients exposed to X-rays for too long can be severely burned or can even die. Patients can also develop problems by getting X-rayed too many times, because X-rays are radioactive. Until the early 1950s, shoe stores used X-ray machines to check the fit of children's feet inside shoes. Children "played" with the machines, taking turns looking at their feet. Once the danger of X-ray radiation became known, these machines immediately disappeared from the stores.

X-rays have many advantages, however. They can be used to destroy cancerous tissue, to expose broken bones or tumors, or to reveal problems like blockages and ulcers. But patients are exposed to X-rays only when necessary.

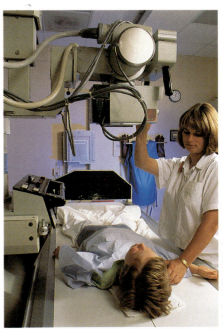

In the top picture, a nurse studies a patient's X-ray. In the bottom picture, a lab technician positions the camera over a child's body in preparation for taking the X-ray itself.

THE ELECTROCARDIOGRAPH (EKG)

The electrocardiograph, or EKG, is a piece of equipment that is used to diagnose heart problems. This extraordinary machine is sometimes called an ECG, but its more common name, EKG, comes from the German version of the word *electrokardiograf*.

The development of the EKG was a direct result of work scientists did in the nineteenth century. Let's take a look at what they did to make this invention possible.

In the 1790s, scientists experimenting with frogs discovered that when electricity was applied to a muscle, that muscle moved. About sixty years later, two German scientists, Rudolph von Kelliker and Herman Muller (Myoo-ler), also did experiments with muscles and electricity.

In 1878, scientists attached electrodes (small bits of metal that conduct electricity) to the hearts of animals and

A technician watches the monitor displaying a patient's EKG readings. These readings tell if a patient's heart is working correctly.

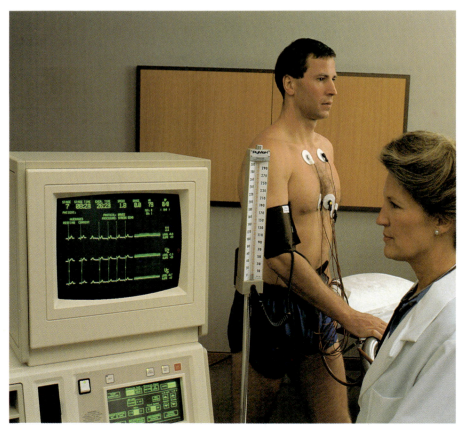

© Jon Feingersh/Tom Stack & Associates

proved that there was electricity in those hearts. In 1887, Augustus Waller proved that there was electricity in the human heart, too.

All of these discoveries led to Willem Einthoven (Ine-toe-ven) measuring electricity in the heart. Einthoven was born in 1860 and moved from the Dutch West Indies to Holland when he was ten. He was a doctor who was also interested in physics and other sciences. In 1903, he built what he called a "string galvanometer." This was a horseshoe-shaped magnet with a wire across its poles. It measured the body's electric current.

Einthoven kept improving this idea until he created the electrocardiograph. The only problem with his invention

Willem Einthoven's string galvanometer measured the body's electrical current. The electrical current passes from the string galvanometer to the body, goes through the body, and then returns to the galvanometer, where it registers. Einthoven improved this machine until he created the electrocardiograph.

was that it weighed eight hundred pounds (360 kg) and was so huge that it couldn't be moved to the hospital where it was to be used. The hospital had to run 1,650 yards (1,500 m) of wire to his lab instead.

Today's machine is much smaller and more improved, of course, but it is still very similar to the one Einthoven first designed. The EKG is used to diagnose heart problems. Doctors place electrodes on the surface of the body, and from the printout (called an electrocardiogram) made from the EKG, they can tell if the heart is working right. If it is not working properly, they can tell just where the problem is. If the patient has had a heart attack, or has any other kind of heart problem, the EKG shows abnormal electric current. The patient can then begin immediate treatment to correct the problem.

For his remarkable and valuable invention, Willem Einthoven won a Nobel prize in 1924.

THE ELECTROENCEPHALOGRAPH (EEG)

Just as the EKG measures and records electrical activity in the heart, the electroencephalograph, or EEG, measures and records electrical activity in the brain.

The brain produces wave patterns that change rhythmically. In 1929, German scientist Hans Berger invented the EEG. To make an electroencephalogram, as the printout is called, doctors place sets of electrodes on the patient's scalp. Each set of electrodes sends a signal back to the EEG showing the differences in voltage between the electrodes. This shows up as a wave pattern on the printout.

A normal person's EEG shows even patterns of what are called *alpha* waves. The pattern of these waves changes when the person's mood changes. For instance, when a person is asleep or in a coma, the waves are slow. When he or she is excited, the waves are low in voltage, fast, and uneven. When a person has brain damage, the damaged area produces slow *delta* waves.

The EEG is very useful for detecting tumors and for studying epilepsy, serious head injuries, and problems with the nervous system. It is limited, however, because it only records a small sample of the brain's electrical activity. Because of this limitation, other machines, such as the MRI and the MEG, have been invented, and they are discussed later in this chapter.

ULTRASOUND

More often today, medical advancements are invented by companies or teams of scientists rather than by individuals. Ultrasound, a new technique for diagnosing disease, is an example of this kind of scientific teamwork.

Invented in the early 1970s, ultrasound uses high frequency sound waves, which hit organs in the body and produce echoes. These echoes display a picture of the organ on a television-like screen.

Radiologists, doctors who are trained to use X-rays and other technology employing light or sound for diagnosis, use ultrasound to find tumors and other masses in the body. Cardiologists, or heart doctors, use it to check that the heart is working properly and that the blood is flowing without interruption. Vascular surgeons, doctors who study blood vessels, use ultrasound to see if anything is blocking the patient's blood vessels. Obstetricians, doctors who specialize in pregnancy and childbirth, use ultrasound to see a growing child inside the mother's womb. By using ultrasound, the obstetrician can sometimes tell if the baby is a boy or a girl.

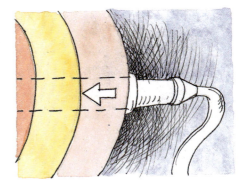

1. The ultrasound machine sends high frequency sound waves to the body.

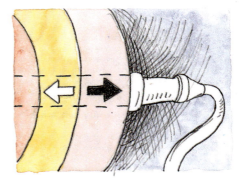

2. The sound waves hit the body and produce echoes.

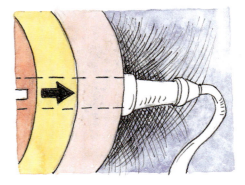

3. The echoes return to the ultrasound machine, creating a picture on a television-like screen.

Dentists also use ultrasound. A jet of liquid along with the high frequency sound waves of ultrasound removes tartar from a patient's teeth.

Optometrists, too, use ultrasound to clean dirt from eyeglasses. The glasses are dipped in water, and sound waves knock off the dirt.

MAGNETIC RESONANCE IMAGING (MRI)

During the 1960s, Dr. Raymond Davadian was convinced that there was a way to look at the chemical makeup of every cell in the human body without using radiation or cutting into the body. Dr. Davadian developed what was originally called "nuclear magnetic resonance," or NMR.

Dr. Davadian knew that cells are made up of atoms and that each atom in the body has a nucleus, or center, which is surrounded by a magnetic field. He wanted to invent a machine that could pick up the magnetic signals from the nucleus. By reading the pattern of the signals, he would

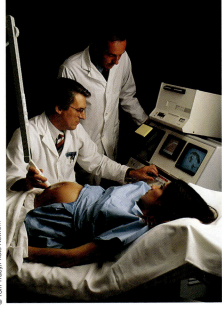

© Tom Tracy/Photo Network

Above: These doctors are watching monitors that display the ultrasound pictures of this woman's baby. If the baby is in the right position, the doctors can tell whether it is a boy or a girl.

At right: A nurse gets a patient ready to slide into the MRI machine, where a part of the body will be scanned for possible problems. Sometimes, the patient must lie perfectly still for forty-five minutes for the scan to be done properly.

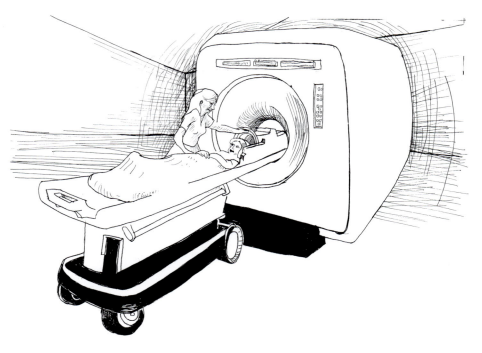

then be able to detect problems. Many doctors didn't believe in his idea, but Davadian built the machine anyway.

The machine is a scanner shaped like a long tube. The patient lies on a table, which slides into the scanner. The MRI bombards the body with strong pulses of radiowaves, which cause the nuclei to rotate, line up, and give off magnetic signals. The scanner picks up these magnetic signals, and a computer displays a very accurate three-dimensional image of the body's tissues on a screen. By reading the screen, doctors can see infections, cancer, spinal and joint problems, brain lesions, and other problems.

THE MAGNETOENCEPHALOGRAPH (MEG)

The magnetoencephalograph, or MEG, is a new development in medical technology. Several scientists wanted to build a machine that would map the brain and the body even more accurately than the MRI, which cannot pinpoint the exact locations of many problems.

A team of scientists from a company called Biomagnetic Technologies, Inc. (BTI), in San Diego, California, built a machine called a "biomagnetometer."

The biomagnetometer takes "biomagnetic" pictures—that is, pictures of the electrical activity in the brain. This is called "biomagnetic imaging."

The patient lies on a nonmagnetic, padded, adjustable table. A sensor is inside a long cylinder called a dewar (*doo-er*) that hangs from the ceiling. The bottom of the cylinder is concave so it fits neatly onto a patient's head.

Among other things, the MEG can identify the location of seizures in epileptic patients, help research the recovery process of trauma and stroke victims, and help detect Alzheimer's disease.

4

Breakthroughs
in Medical
Procedures

When you have appendicitis or tonsillitis, or some other medical problem requiring an operation, you go to the hospital for help. But doctors didn't always know how to perform operations, because often they didn't know what was causing the problem.

From prehistoric days to the days of the ancient Peruvians, people believed disease was caused by evil spirits trapped inside the sick person's head. To rid the person of the disease, they cut a hole in the skull with a short knife to let the evil spirits escape. This practice, called "trephination," had been abandoned by the time the Romans conquered the world. As medical knowledge progressed, more light was shed on the causes of disease.

Many countries made great strides in medical procedures. The Chinese, for example, believed that the universe was made up of two opposing forces, yin and yang. Disease resulted when the two forces were out of balance. In Chinese philosophy and art, yin is the female element, which stands for darkness, cold, and death, while yang is the male element, the source of light, life, and heat. The Chinese developed the art of acupuncture as a way of bringing these forces back into balance. Acupuncture is the practice of putting needles into certain areas of the body. Though it has been practiced for over 2,000 years, it is only now gaining acceptance in the western world, and scientists still do not fully understand how it works.

In India, as long ago as A.D. 700, doctors delivered babies by caesarean section. A caesarean birth is one in which the abdomen and uterus are cut open to remove the baby because a vaginal birth is not possible. (Caesarean sections got their name from Julius Caesar, who is the first known person to have been born in this manner.) The ancient Indians also performed a form of plastic surgery, repairing defective body parts with transplanted healthy tissue.

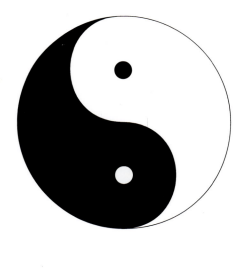

This illustration shows the Chinese symbol of yin and yang. The black side is yin, the female element, and the white side is yang, the male element.

During the Middle Ages (A.D. 400–1400), however, political confusion prevailed and people in Europe lost the many important discoveries of the Romans. They let their cities become filthy with garbage and rats. As a result, bubonic plague and other epidemics routinely killed people. Bubonic plague was a contagious disease transmitted to humans by fleas from infected rats. Victims suffered from high fevers and swollen lymph glands, and often died.

All of the advances made earlier by the ancient Romans, Indians, and Chinese had to wait until those in the western world rediscovered them.

SURGERY BECOMES A SCIENCE

Surprisingly, surgery was not considered a part of medicine until the Middle Ages in Europe were almost over. Doctors left the "lowly" chore of surgery to laborers, like barbers, and some continued that practice until the 1800s.

During the 1400s and 1500s, in Italy and in parts of France, surgery began to gain respectability.

The person credited with advancing surgery to the level of science was not well-educated and wasn't even a doctor. Ambroise Pare (Par-ay), born in 1510 near Laval, France, was a French army barber/surgeon who grew up in a family of barber/surgeons. He soon discovered that he liked surgery more than cutting hair.

Pare learned a great deal about how to remove gunshots. One practice was to burn wounds with hot oil to make them stop bleeding. But one time Pare ran out of oil, so he stitched some of his patients' wounds together instead. When he checked them the next day, he found that their wounds were healing faster than those who had had the hot oil treatment. Use of stitches during amputations was Pare's greatest contribution to surgery.

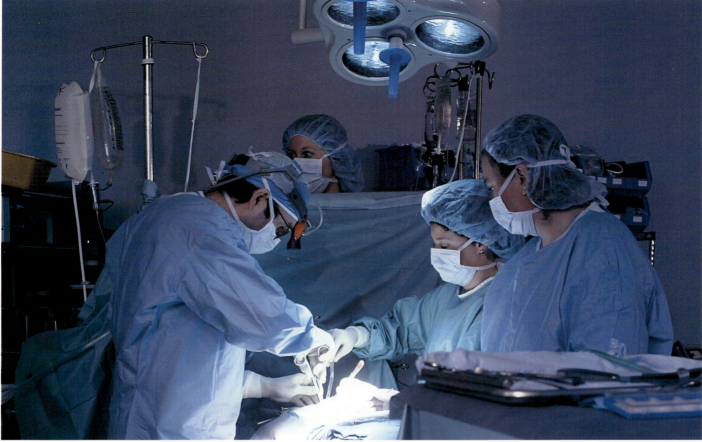

© Al Cook/Photo Network

Pare wrote extensively on surgery and published four editions of his works. In addition, he discovered that changing the position of a baby while it was still in the mother's womb could prevent difficult births. He also invented artificial limbs and eyes. Surgeon to four kings of France, Pare is called the "Father of Modern Surgery."

PASTEURIZATION

If you look at a milk carton in the grocery store, you will see the word "pasteurized" on it. The term comes from the name of the great scientist Louis Pasteur (Pas-*ter*). Pasteur discovered that certain microbes were responsible for the rapid spoiling of milk, cheese, and wine. He found a way to kill these microbes by heating them. This process is called pasteurization, and is still in use today.

A team of four surgeons works on a patient in the operating room. The anesthesiologist stands behind the patient. At one time, surgery was not considered suitable work for doctors and was left to barbers and other unskilled people! Today, of course, surgery is one of the most highly regarded of the medical professions.

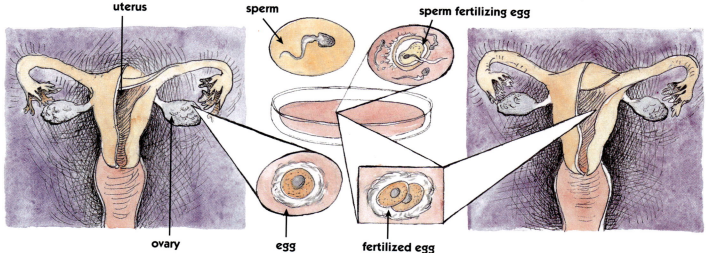

uterus

sperm

sperm fertilizing egg

ovary

egg

fertilized egg

IN VITRO FERTILIZATION

1. In the first step of *in vitro* fertilization, the doctors remove an egg from the woman's ovary.
2. Then the doctors fertilize the egg with the man's sperm in a petri dish in a laboratory.
3. The fertilized egg is then put into the woman's uterus, where it may attach itself and grow into a baby.

In the late 1970s, a new method was developed to help women who had trouble getting pregnant to have a baby. This method is called *in vitro* fertilization. *In vitro* is Latin for the words "in glass."

With this technique, several eggs are removed from the woman's ovaries. Although doctors originally had to operate to remove the eggs, they can now remove them without surgery. Those eggs are then united with the husband's sperm outside of the woman's body. Once an egg is fertilized by a sperm, it is placed inside the woman's uterus, where there is a 15 percent chance that the fertilized egg will grow. If more than one egg is fertilized, they can be frozen for later use.

The first successful *in vitro* fertilization was performed in England in 1977. The baby's name is Louise Brown. Many newspapers and magazines called her the world's first "test-tube" baby. This term is not really accurate. Although Louise was not conceived in the traditional way, she did grow to term in her mother's womb. A true test-tube baby is one who spends the entire nine months prior to birth in an artificial womb. No babies have been born yet in this way. More than 1,000 have been born, however, using the *in vitro* method.

GENETIC ENGINEERING

Every plant and animal on earth is composed of cells. Within each cell are chromosomes, little x-shaped structures containing hereditary material. Humans have forty-six pairs of chromosomes, with twenty-three pairs inherited from each parent.

Each chromosome contains thousands of genes. Each gene carries the code for a specific trait, such as hair color, eye color, height, skin color, intelligence, and so on.

Genes are made of "deoxyribonucleic acid," which is referred to as DNA. DNA is the "genetic code" of life. The study of genes is called genetics. Geneticists are now learning ways to change these codes so that defective genes can be made healthy.

For instance, in 1977, researchers discovered the gene for making insulin. They were able to recreate an artificial form of the gene, which they placed into a bacteria called *Escherichia coli*, or *E. coli*.

E. coli duplicates material, and copies the artificial gene to produce insulin. Diabetics are now using this type of insulin, of which supplies are plentiful.

Scientists have discovered 4,000 genetic problems from which humans can suffer. One is stunted growth, or dwarfism. Through genetic engineering, scientists have been able to produce a growth hormone, which, when given to children, can make them grow properly.

Scientists also discovered the gene that makes Factor VIII, the blood clotting substance most hemophiliacs need to prevent them from bleeding excessively. Factor VIII is becoming more readily available as time goes on.

In the future, doctors hope to be able to transplant healthy genes into people to replace damaged ones that cause certain diseases or abnormalities.

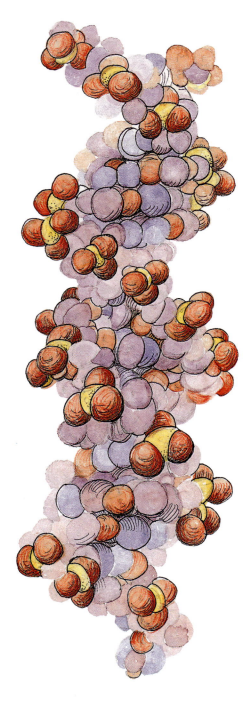

Genes are made of DNA. In 1953, James Watson and Francis Crick discovered that DNA consists of two long strands wound around each other. This shape is called a "double helix." When a baby is created, it receives one strand of DNA from each of its parents.

5

BREAKTHROUGHS IN REPLACEMENT BODY PARTS

In the old television series, *The Six Million Dollar Man,* a pilot who loses an eye and arm and both legs in a terrible plane crash is rehabilitated by receiving "bionic" body parts to replace the ones he lost. His bionic parts look normal but are far more powerful than his original ones.

While bionic parts that make a person stronger than before do not exist, replacement body parts are available for the heart, liver, kidney, and joints. These artificial organs and joints help thousands of people lead normal lives.

PACEMAKER

A pacemaker is an electronic device implanted in a person's chest to regulate the heartbeat. Pacemaker technology evolved as a direct spinoff of the space program.

Some people are unable to maintain a normal heart rhythm due to heart disease. Pacemakers were initially developed to overcome heart block, a condition where the body's electric signals, which trigger the heart to beat, fail to carry their message. Rather than beating regularly, the heart skips, and becomes slow and irregular. This can cause cardiac arrest, which can be fatal. Before pacemakers, these people would have died. With pacemakers, however, they can lead normal active lives.

Two men are credited with inventing the pacemaker. Dr. Paul Zoll invented an external pacemaker in the 1930s. Called the "NTP" (Noninvasive [without surgery] Temporary Pacemaker), an external pacemaker is one in which the power supply is located outside the body. Dr. C. Walton Lillihei is also credited with the discovery of the pacemaker. He designed one in 1957, which was built by a television repairman, Earl Bakken, in just one month.

Dr. Lillihei's pacemaker wasn't ready to be implanted in a human, however, because it needed to be made smaller

Courtesy of Medtronic, Inc.

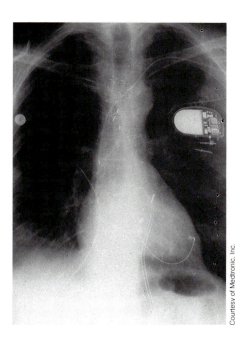

Courtesy of Medtronic, Inc.

The top picture shows an internal pacemaker, which is only a little more than a few centimeters wide. Because it is so small, the patient cannot even feel it. The bottom picture shows an internal pacemaker implanted in the chest of a patient. An electrode attaches to the heart, and a power supply provides electric signals to the heart to help regulate the heartbeat.

and to be wrapped in plastic. Earl Bakken decided to start a company called Medtronics to make pacemakers. Today, Medtronics is the largest supplier of pacemakers.

Though external pacemakers were the earliest form, they are still used during the complicated procedure of open heart surgery and after bypass operations to regulate the heartbeat. Today, pacemakers are entirely implanted in the body and they work with the heart's natural rhythm. If the heart is beating normally, the pacemaker does nothing more than monitor it. If, however, the heartbeat is disturbed, the pacemaker begins to regulate it.

ARTIFICIAL JOINTS

A joint is a point on the body where moving parts are joined. The knees, hips, and shoulders are all joints. When people have problems with their joints, movement can become difficult, painful, or even impossible. Many people in the world suffer from crippling arthritis and other diseases that affect joints, and as our population ages, this figure is expected to rise.

In the past, many injured people were doomed to spend their lives on crutches or in wheelchairs, but thanks to new technology many injuries can now be repaired by artificial joints.

A British orthopedic surgeon, Sir John Charney, pioneered artificial joint technology with hip replacements in the 1960s. The most common joints replaced today are the hip and knee, but feet, shoulders, elbows, and hands can also be replaced. Wrists, ankles, and toes are rarely replaced; however, ankles have been successfully fused to prevent arthritic pain.

The first artificial hip joints were made of metals like chrome-cobalt alloys or stainless steel. Now they are made

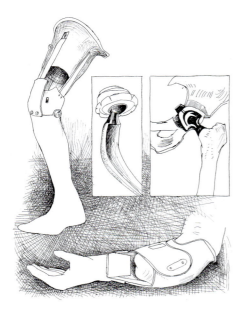

Today, scientists can build artificial joints, including those for the knees, shoulders, hips, and elbows. These artificial joints help injured people lead normal lives.

of titanium alloys, since these match the elasticity of bone more closely. In many cases, the artificial joint will last a lifetime.

Artificial joints are not exactly like real body parts, but their introduction has changed the course of medical science and altered many people's lives for the better.

ORGAN TRANSPLANTS

Hearts, livers, and kidneys are the most commonly transplanted organs today. Bone marrow is also often transplanted.

On December 3, 1967, Dr. Christian Barnard, a doctor at Groote Schoor (Groh-tuh Shoor) Hospital in South Africa, stunned the world by announcing that he had performed a heart transplant. He transplanted the heart of a twenty-five-year-old woman who had died in a car accident into the body of fifty-five-year-old Louis Washansky. The new heart worked fine, but Washansky died of other complications eighteen days after the operation.

Since that day, hundreds of heart transplants have been performed. Sometimes patients do not need a whole heart, but just a heart valve. In these cases, doctors have discovered that pigs' heart valves work extremely well in humans.

During the early days of transplants, the patient's body would often reject the new organ as foreign tissue, causing major problems. But in the early 1980s, a new drug, cyclosporine, became available and was very effective in eliminating the rejection problem.

Often, human body parts for transplants are hard to find, so scientists are constantly researching ways to create artificial ones. But doctors and patients must currently rely on the generosity of human donors who give permission for their organs to be used in the event of their death.

BIBLIOGRAPHY

Bedeschi, Giulio. *Science of Medicine.* New York: Collins Publishers, Franklin Watts, Inc., 1975.

Bender, George A. *Great Moments in Medicine.* Detroit: Parke-Davis, 1961.

Bowen, Robert Sidney. *They Found the Unknown.* Philadelphia: MacRae Smith Company, 1963.

Calder, Ritchie. *The Wonderful World of Medicine.* New York: Doubleday and Company, Inc., 1969.

Camp, John. *Magic, Myth, and Medicine.* New York: Taplinger Publishing Co., Inc., 1973.

Clark, John, ed. *The Human Body: The Cell: A Small Wonder.* New York: Torstar Books, Inc., 1985.

Cross, Wilbur and Susan Graves. *The New Age of Medical Discovery.* New York: Hawthorn Books, Inc., 1972.

Dietz, David. *All About Great Medical Discoveries.* New York: Random House, 1960.

Dolan, Edward F., Jr. *Inventors for Medicine.* New York: Crown Publishers, Inc., 1971.

Duffy, John. *The Healers: The Rise of the Medical Establishment.* New York: McGraw-Hill Book Company, 1976.

Fradin, David Brindell. *Medicine: Yesterday, Today, and Tomorrow.* Chicago: Children's Press, 1989.

Goldstein, Kenneth K. *New Frontiers of Medicine.* Boston: Little, Brown and Company, 1974.

Hume, Ruth Fox. *Great Men of Medicine.* New York: Random House, 1961.

Marti-Ibanez, Felix. *History of American Medicine.* New York: MD Publications, Inc., 1959.

Morse, Joseph Laffan. *Funk and Wagnalls New Encyclopedia.* Volumes 8, 9, 12, 15, 16, 18, and 21. New York: Funk and Wagnalls, Inc., 1973.

Shorter, Edward, Ph.D. *The Health Century.* New York: Doubleday, 1987.

Williams, Guy. *The Age of Agony.* Chicago: Academy Chicago Publishers, 1986.

van Zandt, Eleanor. *Twenty Names in Medicine.* New York: Marshall Cavendish, 1988.

INDEX